Guide to Infectious Diseases by Body System

Jeffrey C. Pommerville
Glendale Community College

Introduction

For many students entering the allied health or nursing fields, connecting both microbiology and anatomy and physiology in terms of infectious disease can be a very rewarding and necessary experience. Therefore, I have put together this short guide to infectious diseases according to the anatomical body systems they most commonly affect. Since some diseases affect more than one body system, I have tried to match diseases with the more likely affected body system, but some flexibility is called for when you encounter these diseases.

Each "anatomical unit" begins with **Background** material and then identifies the major **Signs and Symptoms** for the diseases of that body system. Each unit further considers some **Common Clinical Conditions** for these diseases. Each unit ends with a brief synopsis of drug **Treatments**.

I hope you find this guide useful in your microbiology studies and health careers.

Jeffrey Pommerville

Bacterial and Fungal Skin Infections

Background

The skin is the largest organ of the human body and provides an effective mechanical and chemical barrier to infectious agents. Therefore, it is not surprising that many skin infections and disease are caused by opportunistic pathogens that reside as part of the normal human microbiota.

Most skin diseases result from scrapes or punctures that break the skin. **Folliculitis,** an infection of the hair follicles, **cellulitis,** an acute inflammation of the skin and subcutaneous tissues, and a few other diseases can result from direct contact with several species of bacteria. The most common species are *Staphylococcus aureus* and *Streptococcus pyogenes* (group A). Since the infection site often contains pus, such bacteria are termed **pyogenic** (pus-producing).

Several species of fungi (dermatophytes) can infect the outer layer of the epidermis, causing **cutaneous mycoses,** commonly called **ringworm**.

The entry or infection of a pathogen into the skin epithelium often results in some form of lesion, including:

Macules: Flat spots on the skin that often are reddened due to inflammation.

Pustules: Raised, red spots on the skin containing pus; without pus, they are called **papules**.

Vesicles: Small, raised skin lesions that contain fluid.

Bullae: Larger skin lesions containing liquid.

Signs and Symptoms

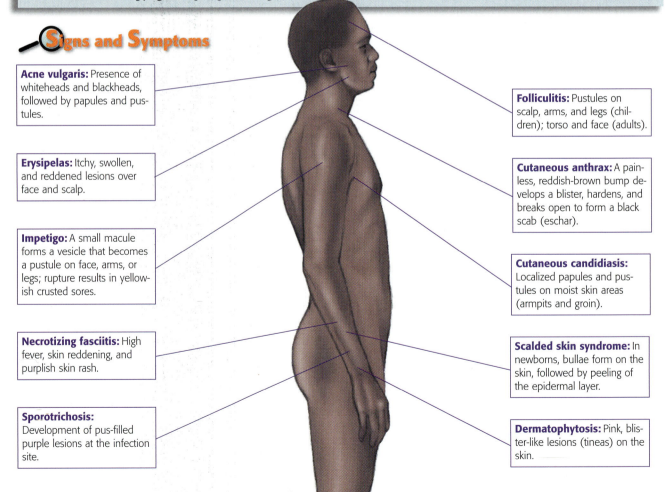

Acne vulgaris: Presence of whiteheads and blackheads, followed by papules and pustules.

Erysipelas: Itchy, swollen, and reddened lesions over face and scalp.

Impetigo: A small macule forms a vesicle that becomes a pustule on face, arms, or legs; rupture results in yellowish crusted sores.

Necrotizing fasciitis: High fever, skin reddening, and purplish skin rash.

Sporotrichosis: Development of pus-filled purple lesions at the infection site.

Folliculitis: Pustules on scalp, arms, and legs (children); torso and face (adults).

Cutaneous anthrax: A painless, reddish-brown bump develops a blister, hardens, and breaks open to form a black scab (eschar).

Cutaneous candidiasis: Localized papules and pustules on moist skin areas (armpits and groin).

Scalded skin syndrome: In newborns, bullae form on the skin, followed by peeling of the epidermal layer.

Dermatophytosis: Pink, blister-like lesions (tineas) on the skin.

Common Clinical Conditions

The following is a brief description of several bacterial and fungal diseases of the skin.

Bacterial Skin Diseases

Acne vulgaris: A common, localized inflammation of hair follicles on the face and upper torso of adolescents. Sex hormone changes and blocked sebaceous ducts cause rupture of the hair follicle lining, triggering infection by *Propionibacterium acnes.* Severe cases can leave permanent scars on the skin.

Erysipelas: An inflammation resulting from the entry of *S. pyogenes* through cracks in the dermis. The disease can be life threatening without treatment.

Necrotizing fasciitis: A very severe, deep cutaneous infection caused by *S. pyogenes.* Blood vessel clotting results in extensive destruction of fat and muscle tissue. Prompt therapy is needed to prevent systemic spread.

Cutaneous anthrax: A serious disease caused by spores of *Bacillus anthracis* that enter through a break in the skin. Anthrax bacilli produce three exotoxins that have the potential to kill about 20% of exposed and untreated individuals.

Folliculitis: An infection of hair follicles by *S. aureus.* A superficial infection involves a single hair follicle and a deep infection of the follicle forms a **furuncle** (boil). Abscesses of several interconnecting furuncles form a **carbuncle.**

Scalded skin syndrome: An *S. aureus* infection prevalent in newborns and young children. A bacterial toxin causes the epidermis to separate from the dermis, causing blistering, reddening, and peeling of the top layers. Adults usually are resistant to infection.

Impetigo: A form of cellulitis that usually occurs in newborns and children from infection by *S. aureus* or *S. pyogenes.* Although rarely serious, the infection is highly communicable.

Fungal Skin Diseases

Cutaneous candidiasis: A skin or mucosal infection usually caused by *Candida albicans* where excessive moisture exists. By multiplying in the cutaneous skin layers, it triggers the development of local skin lesions in the armpits and groin. Candidiasis of the mouth in newborns is called **thrush.**

Dermatophytosis: Superficial infections, caused by the genera *Epidermophyton, Microsporum,* and *Trichophyton,* are termed **tineas.** A second word indicates the area of infection, such as tinea pedis: foot (athlete's foot) and tinea cruris: groin (jock itch).

Sporotrichosis: A chronic disease caused by *Sporothrix schenckii.* The most common form involves small, subcutaneous lesions that may spread and form secondary lesions along the lymphatic system that drain the original lesion.

Treatment

Treatment of bacterial skin diseases is usually successful using specific antibiotics. Ciprofloxacin is useful for cutaneous anthrax. Antifungal compounds, such as tolnaftate, miconazole, and clotrimazole, are used for ringworm and candidiasis (also nystatin). Sporotrichosis can be treated with an oral solution of potassium iodide.

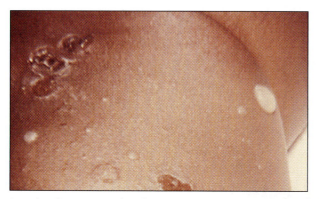

Impetigo. (Courtesy CDC/PHIL)

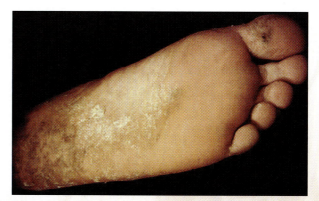

Athlete's foot. (Courtesy CDC/PHIL)

Viral and Parasitic Skin Infections

Background

There are several viruses and a parasite that cause skin infections and disease in humans, and whose clinical symptoms are commonly diagnosed by their effects on the skin. A few are found solely within the skin, while most skin diseases are systemic, spreading from another body system, such as the blood or respiratory system, before appearing as skin lesions. Several of these viral diseases cause a skin rash, called an **exanthema**. These include chickenpox, measles, and rubella.

Viruses that cause skin diseases belong to one of four groups based on genome architecture:

Double-stranded (ds) DNA Viruses
Papovaviridae: Among these viruses are some 80 human papillomaviruses that can cause warts.

Poxviridae: This group includes the viruses responsible for smallpox, monkeypox, and molluscum contagiosum.

Herpesviridae: These viruses include: varicella-zoster virus (VZV) that causes chickenpox and shingles; herpes simplex virus 1 (HSV-1) that causes cold sores; and human herpesvirus 6 (HHV-6) that causes roseola.

Single-stranded (ss) DNA viruses
Parvoviridae: Within this group is parvovirus B19 that causes erythema infectiosum.

ssRNA Viruses (– strand)
Paramyxoviridae: Among these viruses, the rubeola virus is the etiologic agent for measles.

ssRNA Viruses (+ strand)
Togaviridae: This group includes the rubella virus that causes German measles.

Signs and Symptoms

Measles: In children, fever, malaise, cough with rash on upper body.

Roseola: High fever followed by a profuse maculopapular rash on the arms, neck, and trunk with accompanying drop in body temperature.

Chickenpox: In children, fever, maculopapular rash, and itching on face and trunk.
Shingles: In adults, reddish blistering on trunk and back with intense, debilitating pain.

Smallpox: Fever, followed by prominent vesicles on face, arms, and legs that become papules and then pustules.

Rubella: In children, maculopapular red rash and low-grade fever.

Monkeypox: Fever, headache, muscle aches, and backache; followed by vesicular rash spreading from face to body.

Herpes simplex: Lesions (cold sores/fever blisters) along the margins of the mouth (lips).

Erythema infectiosum: In children, low grade fever followed by a red, macular rash on arms and face (slapped cheek appearance) that spreads to body.

Molluscum contagiosum: Small, waxy raised papules on the face and eyelids of children, and in adults on breasts, genitalia, and inner thigh.

Cutaneous leishmaniasis: One or more papules, leaving a flat, atrophic scar on healing.

Warts: Small, elevated, usually benign skin growths typically on hands and soles of feet.

Common Clinical Conditions

The following is a brief description of several viral and parasitic diseases of the skin.

Viral Skin Diseases

Herpes simplex: In a primary infection, HSV-1 may produce blisters. Following a latent stage, stress or trauma triggers the development of vesicles (**cold sores/fever blisters**) that are infectious. The blisters heal as the infection subsides.

Roseola: An acute but benign infection in infants and young children. Sixty percent of cases are inapparent infections and involve only a fever.

Chickenpox (varicella): A mild, but very contagious disease of childhood. After passage through the respiratory system, VZV infects skin cells forming vesicles. The vesicles/pustules rupture and scab over during healing.

Shingles (herpes zoster): Latent VZV becomes activated often after decades of dormancy. Vesicular lesions typically form on the waist and cause severe pain.

Measles: The virus is spread by the respiratory route, making measles a very contagious disease among school-age children. Symptoms start with a sore throat, coughing, and headache. Tiny red lesions with white centers in the mouth (**Koplik spots**) are a diagnostic key.

Rubella: Although highly contagious via respiratory droplets, rubella (**German measles**) is much milder than measles. The rubella virus typically infects children.

Erythema infectiosum: A contagious disease (also called **fifth disease**) primarily in children. It is caused by human parvovirus B19. Epidemics sometimes affect the adult population.

Smallpox: The variola virus is highly communicable and is transmitted via the respiratory route through the bloodstream to the skin. The vesicles and pitted lesions (**pox**) that develop are less prominent on the trunk.

Monkeypox: A rare viral disease whose first appearance in the Western Hemisphere occurred in the Midwest in 2003. It is much milder and less contagious than smallpox.

Molluscum contagiosum: A mildly contagious but uncommon disease spread by direct or indirect contact. The virus, which can infect any part of the skin, mainly affects children and young adults.

Warts (verrucae): A papillomavirus infection on the surface of mucous membranes or dead layers of the skin. Infection stimulates excessive epithelial cell divisions usually resulting in benign tumors (**common warts** or **plantar warts**).

Parasitic Skin Diseases

Cutaneous leishmaniasis: This is the most common form of the disease. It arises from the bite of a sandfly (*Phlebotomus*) infected with the protozoan *Leishmania*.

Treatment

Antibiotics are ineffective against viral infections. Diseases like chickenpox, measles, and rubella are overcome by the immune system. Vaccines also are available. Shingles can be treated with the antiviral drug acyclovir to alleviate symptoms. The drug also is useful in preventing lesion spread and shortening healing time for cold sores.

Warts sometimes spontaneously disappear. They can be removed with cryotherapy (extremely cold liquid nitrogen) or through "tissue burning" with caustic chemical agents. Vaccination has eradicated smallpox worldwide. An antimony compound is effective against cutaneous leishmaniasis

Leishmania donovani

Chickenpox. (Courtesy CDC/PHIL)

Eye Infections

Background

Being exposed to the environment, the eyes represent a portal of entry for infection and disease. In fact, the epithelium that covers the eyes is a continuation of the skin and mucosa. Normally, eye infections are rare because (1) several microbes of the human microbiota, especially *Staphylococcus epidermidis* and *Corynebacterium xerosis,* normally colonize the surface of the eyes, and (2) the presence of secretory IgA antibody and lysozyme in tears continually flush the eyes. Scratches or very dry eyes can decrease the effectiveness of these protective mechanisms and allow infection to occur.

Several bacteria, viruses, and parasites are associated with eye infections. The diseases fall into three categories. **Infectious conjunctivitis** is an inflammation of the conjunctiva, which is the thin mucous membrane that lines the inner surface of the eyelids and the outer surface of the eyeball. **Keratitis** is an inflammation of the cornea, the transparent part of the eyeball covering the iris and the lens. Extensive disease involving both the conjunctiva and the cornea is called **keratoconjunctivitis.**

The etiologic agents of eye infections include:

Bacteria:	*Streptococcus pneumoniae*
	Staphylococcus aureus
	Haemophilus influenzae (biogroup aegyptius)
	Neisseria gonorrhoeae
	Chlamydia trachomatis
Viruses:	Specific serotypes of the *Adenoviridae*
	Herpes simplex virus 1 (HSV-1)
Parasites:	*Acanthamoeba*
	Onchocerca volvulus

Signs and Symptoms

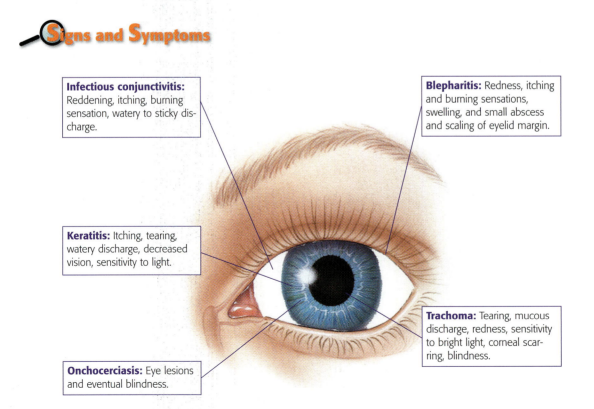

Infectious conjunctivitis: Reddening, itching, burning sensation, watery to sticky discharge.

Blepharitis: Redness, itching and burning sensations, swelling, and small abscess and scaling of eyelid margin.

Keratitis: Itching, tearing, watery discharge, decreased vision, sensitivity to light.

Trachoma: Tearing, mucous discharge, redness, sensitivity to bright light, corneal scarring, blindness.

Onchocerciasis: Eye lesions and eventual blindness.

Common Clinical Conditions

The following is a brief description of eye diseases caused by the most common microbial agents.

Blepharitis: This can involve an acute or chronic inflammation often caused by *S. aureus* at the base of an eyelash (**stye**) or blocking of a sebaceous gland on the eyelid.

Infectious conjunctivitis

Bacterial (acute): Primarily caused by *H. influenzae* or *S. pneumoniae,* this form of conjunctivitis is commonly called **pinkeye**. Usually, there is no affect on vision.

Neonatal: Gonorrheal ophthalmia, caused by *N. gonorrhoeae,* and **chlamydial ophthalmia,** caused by specific serovars of *C. trachomatis,* are serious forms of acute conjunctivitis. Infection occurs as the newborn passes through the birth canal of a mother having gonorrhea or a chlamydial infection.

Adult inclusion: Caused by similar serovars of *C. trachomatis,* this is mild in form and rarely causes blindness. It is spread from genitalia to the eye.

Epidemic conjunctivitis (keratoconjunctivitis): Commonly caused by specific *Adenoviridae* serotypes. Also often referred to as **pinkeye,** it can be extremely contagious but resolves without treatment.

Keratitis

- The protozoan *Acanthamoeba* causes a form of keratitis that has become more prevalent with the wearing of contact lenses that are not properly disinfected. Outbreaks also have been associated with improperly chlorinated swimming pools, communal sharing of infected towels, or ophthalmic procedures using contaminated equipment.

- HSV-1, which also causes cold sores (fever blisters), is a major cause of adult eye disease (**herpetic keratitis**). Inflammation starts in the conjunctiva but eventually affects the cornea, causing corneal damage.

Trachoma: A chronic form of keratoconjunctivitis, it is caused by a different set of *C. trachomatis* serovars that cause inflammation of the conjunctiva and eventual corneal scarring. Blindness can result. Found primarily in developing nations, it is the greatest single infectious cause of adult blindness worldwide. Transmission is by direct or indirect contact and reinfection is common.

Onchocerciasis: This skin and eye disease in tropical Africa, Mexico, and Central and South America, commonly called **river blindness,** is caused by *O. volvulus.* Carried by black flies (genus *Simulium*), the helminth can infect many body tissues, the worst affect being larval worms (microfilariae) that can migrate to the eyes, especially the cornea. The death of microfilariae is toxic to the eye and over several years can lead to blindness. Onchocerciasis ranks second to trachoma as an infectious cause of blindness.

Treatment

Neonatal gonorrhea can be treated with the antibiotic ceftriaxone, while erythromycin eye drops prevent the disease in neonates. Topically-applied sulfonamide ointment is an effective treatment for conjunctivitis. Although herpetic keratitis usually resolves on its own, infections can be effectively treated with oral acyclovir or trifluridine eye drops.

Although *C. trachomatis* is sensitive to several antibiotics (erythromycin or tetracycline), these drugs only control cell growth; they cannot eliminate the bacterium from the eye. Once permanent scarring has occurred, a corneal transplant is necessary to restore vision. Ivermectin works well against *O. volvulus.*

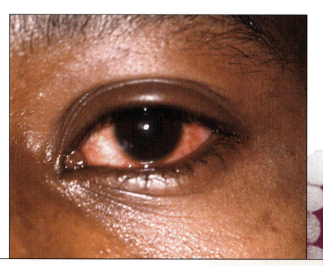

Pinkeye. (Courtesy CDC/PHIL)

A scanning electron micrograph of *S. aureus* cells. (Courtesy CDC)

Bacterial and Viral Infections

Background

The upper respiratory tract (URT) consists of the nose, pharynx (throat), and the middle ear and auditory tubes. The normal microbiota of the URT suppress colonization and inhibit pathogen growth while mucous membranes lining the nose and upper throat trap microbes. Still, URT infections are quite common, and often are transmitted by airborne droplets or droplet nuclei from infected individuals or carriers. About 90% of URT infections are viral, mostly the result of cold viruses.

URT infections exhibit typical clinical **syndromes** (see opposite page).

Pharyngitis: An infection of the pharynx, causing sore throat and sometimes fever. Infection of the tonsils is **tonsillitis**.

Sinusitis: An inflammation of the sinuses causes these areas to fill with fluid and become infected.

Rhinitis: An inflammation and swelling of the nasal mucous membranes.

Otitis media: An infection of the middle ear, often resulting from a nose or throat infection.

Epiglottitis: A bacterial inflammation of the epiglottis (the small flap of skin covering the larynx); swelling can lead to respiratory blockage.

Signs and Symptoms

Otitis media: Deep ear pain, impaired hearing, and bulging eardrum; fever, dizziness, and nausea.

Diphtheria: Moderate fever and sore throat; lesions on pharynx, larynx, and tonsils; grayish pseudomembrane on throat and nasopharynx.

Scarlet fever: High fever, swollen lymph nodes, sore throat; strawberry/raspberry appearance of tongue followed by a diffuse red rash on neck, torso, and extremities.

Strep throat: Sore throat, beefy red pharynx, malaise, fever, and headache; swollen pharynx and lymph nodes.

Common cold: Runny nose, headache, congestion, sneezing, sore throat.

Parainfluenza: Fever, profuse nasal discharge, sore throat, and dry, barking cough.

Other URT Infections

The bacteria responsible for meningococcal meningitis (*Neisseria meningitidis*) and *Haemophilus* meningitis (*Haemophilus influenzae*) can enter by way of the URT. Both will be described with the diseases of the nervous system.

Common Clinical Conditions

Diseases of the URT involve bacteria and viruses.

Bacterial Diseases

Strep throat (streptococcal pharyngitis): This most common streptococcal infection by *S. pyogenes* is transmitted person-to-person via droplets or nasal secretions. Besides an inflamed pharynx, **rheumatic fever** or **scarlet fever** may develop.

Scarlet fever: An initially local tonsillar or pharyngeal infection, the disease can spread as a body rash after the release of exotoxins. The disease is more common in over-crowded areas.

Otitis media: A bacterial or viral infection of the middle ear primarily by *Streptococcus pneumoniae* or *Haemophilus influenzae*. Most common in young children, the infection often results from refluxing through the auditory tubes.

Diphtheria: A highly contagious and dangerous infection of the pharynx mucous membranes (pharyngitis) caused by *Corynebacterium diphtheriae*. Virulent strains produce an exotoxin that damages the throat, causing throat and lymph node swelling. Throat swelling can block the airway leading to suffocation.

Viral Diseases

Common cold (coryza): There are more than 200 different virus subtypes that can cause a common cold, which is an infection of the nose, sinuses, and pharynx. Sinusitis and otitis media may be complications. Signs and symptoms can vary somewhat depending on the infecting virus.

- **Adenoviruses (*Adenoviridae*):** Common colds arising from an adenovirus infection produce a substantial fever and an intense sore throat and cough.
- **Rhinoviruses (*Picornaviridae*):** More than 100 subtypes make the rhinoviruses the most common cause of typical "**head colds**." Signs and symptoms include headache, chills, a scratchy throat, and a runny nose.

Parainfluenza (*Paramyxoviridae*): Human parainfluenza virus subtype 1 (HPIV-1) is the leading cause of **croup** in children. Subtypes 1–3 also cause other upper respiratory and lower respiratory infections. All are much milder than influenza.

Treatment

Although strep throat itself is not usually dangerous, complications of rheumatic fever and kidney damage can be, so treatment with penicillin or erythromycin may be necessary. Amoxicillin can be used to treat otitis media.

Prevention of diphtheria is accomplished through infant immunization with the DPT vaccine. Treatment involves administration of diphtheria antitoxin and antibiotic (penicillin or erythromycin) therapy.

Since antibiotics are useless against viruses, common colds must just run their course. Over-the-counter (nonprescription) medications, including antihistamines and decongestants, may relieve symptoms.

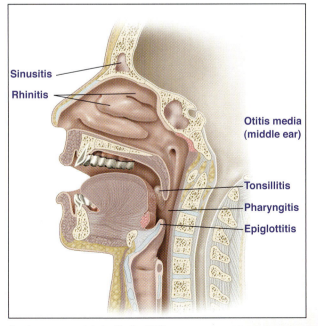

Sinusitis
Rhinitis
Otitis media (middle ear)
Tonsillitis
Pharyngitis
Epiglottitis

Syndromes associated with the URT.

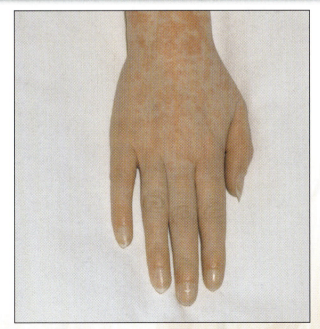

Scarlet fever rash on the wrist and hand. (© Mediscan/Visuals Unlimited)

Bacterial Infections

Background

The lower respiratory tract (LRT) consists of the larynx (voice box), trachea (windpipe), bronchial tubes, and the alveoli. Due to the mucous membrane and filtering mechanisms of the bronchial tubes, the LRT normally contains few microbes. Therefore, if pathogens enter the LRT, serious respiratory disease may result.

Several microbial diseases are associated with the LRT. Two of the bacterial diseases are pneumonia and tuberculosis, historically the two greatest infectious killers of humans. Infections of the LRT also exhibit themselves in several ways (see opposite page):

Bronchitis: An inflammation of the bronchial lining (wall) by viruses or bacteria (*Streptococcus, Mycoplasma, Chlamydia*), producing a thick mucus that narrows the airways.

Bronchiolitis: Usually restricted to young children, a viral infection of the bronchiole lining causes a swelling and narrowing of the airways, making expiration difficult (a wheezing sound heard).

Pneumonia: An acute and complex syndrome resulting from an infection of the lung tissue and alveoli. Impaired gas exchange causes rapid and labored breathing, and cough.

Signs and Symptoms

Inhalation anthrax: Fever, chills, cough, chest pain, headache, malaise; severe breathing and shock result.

Tuberculosis: Fever, fatigue, weight loss, cough; shortness of breath and chest pain; tubercle development.

Primary atypical and chlamydial pneumonia: Gradual and mild symptoms with fever, fatigue, and dry, hacking cough.

Pertussis: Catarrhal stage: malaise, dry cough, fever; Paroxysmal stage: violent (whooping) cough; Convalescent stage: sporadic cough that slowly subsides.

Pneumococcal pneumonia: High fever, chest pain, persistent cough, rust-colored sputum; increased pulse, and difficulty breathing.

Q fever: Dry cough, high fever, chest pain, and severe headache.

Legionellosis: Pneumonia symptoms with fever, dry cough, diarrhea, and vomiting.

Ornithosis: Fever, headache, and dry cough.

Other Pneumonias

Klebsiella pneumoniae and *Serratia marcescens* may produce pneumonias through a hospital-acquired infection.

Common Clinical Conditions

Bacterial diseases of the LRT are outlined below.

Bacterial Pneumonias

Pneumococcal pneumonia: *Streptococcus pneumoniae* is responsible for about 80% of all pneumonia cases. It usually starts after an URT viral infection damages the airways. Without appropriate antibiotic treatment, mortality is high, especially in the elderly.

Primary atypical (walking) pneumonia: Caused by *Mycoplasma pneumoniae,* the infection is common in children and teenagers. The disease is rarely fatal.

Legionellosis (Legionnaires' disease): *Legionella pneumophila* is inhaled as aerosols from air-conditioning devices or water supplies contaminated with the bacteria. After several days of incubation, symptoms appear with pneumonia being the most likely outcome.

Q fever: This pneumonia-like infection, caused by *Coxiella burnetii,* is transmitted by inhaling aerosol droplets or consuming contaminated meat or unpasteurized milk from infected animals. The mortality rate is low.

Ornithosis (psittacosis): A rare pneumonia caused by the bacterium *Chlamydia psittaci.* The obligate intracellular bacteria are inhaled in dried droppings from infected birds (parrots, parakeets, pigeons, turkeys). Most cases are mild.

Chlamydial pneumonia: *Chlamydia pneumoniae* also causes a form of pneumonia with symptoms and outcomes similar to primary atypical pneumonia.

Other Bacterial Diseases

Pertussis (whooping cough): Caused by *Bordetella pertussis,* this highly contagious childhood disease produces mucus in the respiratory system, which triggers coughing. Straining for air causes the "whooping" sound.

Tuberculosis: An infection by *Mycobacterium tuberculosis,* the major causative agent of tuberculosis (TB), starts by inhaling bacilli from an infected person. In the alveoli, the bacilli reproduce, leading to calcified aggregations of activated macrophages and lymphocytes (**tubercles**) surrounding the bacteria.

Inhalation anthrax: Without treatment, this deadly disease begins with typical cold symptoms, but quickly leads to breathing difficulties and shock from toxins produced by the cells from the germinated *Bacillus anthracis* spores.

☤ Treatment

Penicillin is the drug of choice for pneumococcal pneumonia. Primary atypical and chlamydial pneumonia are treated with erythromycin or tetracycline. Legionellosis also can be treated with erythromycin. Q fever can be treated with doxycycline, while ornithosis is best treated with tetracycline.

TB treatment involves extended use of isoniazid and rifampin for 6 to 9 months. Pertussis can be treated with erythromycin and, if caught very early, inhalation anthrax is treatable with ciprofloxacin.

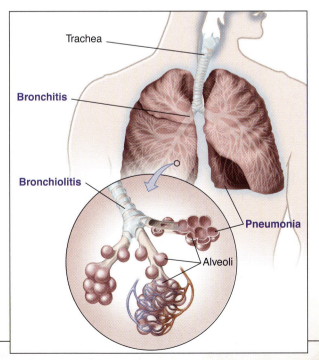

Syndromes associated with the LRT.

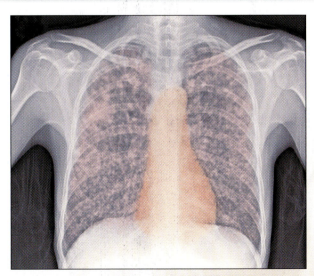

False-colored chest X ray showing tuberculosis. The lungs contain lesions (pink) of infected tissue. (© Du Cane Medical Imaging, Ltd./Photo Researchers, Inc.)

Viral and Fungal Infections

Background

Viral infections of the lower respiratory tract (LRT) include influenza, which kills thousands of people worldwide each year, hantavirus pulmonary syndrome (HPS), and more recently severe acute respiratory syndrome (SARS). Viruses also cause **viral pneumonias,** the most common virus being the respiratory syncytial virus (RSV).

Many of the serious fungal diseases of the LRT represent **systemic mycoses.** The organisms, *Histoplasma capsulatum, Blastomyces dermatitidis,* and *Coccidioides immitis,* primarily infect the lungs. Although most infections are asymptomatic and individuals recover

without antimicrobial therapy, sometimes these fungi spread to secondary sites, causing disease that is more serious. Each type of mycosis is restricted to a specific geographical region. Also, these three fungi are dimorphic, going through a saprobic and parasitic phase (see opposite page).

The saprobic phase of all three fungi grow in the soil where they fragment or form **conidia.** Airborne transmission of these conidia into the respiratory tract triggers the parasitic phase, which can lead to **spherule** development or the formation of **yeast-like forms.**

Signs and Symptoms

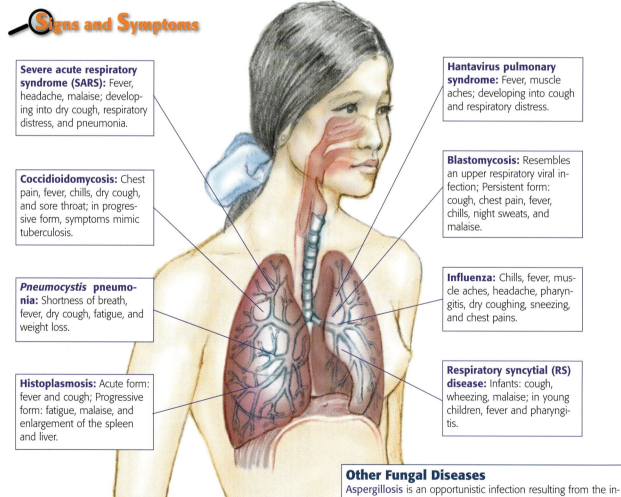

Severe acute respiratory syndrome (SARS): Fever, headache, malaise; developing into dry cough, respiratory distress, and pneumonia.

Coccidioidomycosis: Chest pain, fever, chills, dry cough, and sore throat; in progressive form, symptoms mimic tuberculosis.

***Pneumocystis* pneumonia:** Shortness of breath, fever, dry cough, fatigue, and weight loss.

Histoplasmosis: Acute form: fever and cough; Progressive form: fatigue, malaise, and enlargement of the spleen and liver.

Hantavirus pulmonary syndrome: Fever, muscle aches; developing into cough and respiratory distress.

Blastomycosis: Resembles an upper respiratory viral infection; Persistent form: cough, chest pain, fever, chills, night sweats, and malaise.

Influenza: Chills, fever, muscle aches, headache, pharyngitis, dry coughing, sneezing, and chest pains.

Respiratory syncytial (RS) disease: Infants: cough, wheezing, malaise; in young children, fever and pharyngitis.

Other Fungal Diseases
Aspergillosis is an opportunistic infection resulting from the inhalation of *Aspergillus fumigatus* conidia. Germination of the conidia produces a fungus ball (aspergilloma) in the lungs.

Common Clinical Conditions

Viral and fungal diseases of the LRT are summarized below.

Viral Diseases

Influenza (flu): The influenza viruses are members of the *Orthomyxoviridae*. Transmission is by direct or indirect contact. The virus infects the bronchi and bronchioles. Illness is often severe with symptoms lasting for 3 to 5 days.

Respiratory syncytial (RS) disease: The respiratory syncytial virus (RSV) is a member of the *Paramyxoviridae* and represents the most common cause of bronchitis, bronchiolitis, and viral pneumonia among infants and young children. The virus is transmitted by respiratory secretions.

Severe acute respiratory syndrome (SARS): This is an emerging infectious disease caused by a member of the *Coronaviridae*; 10–20% of affected individuals require mechanical ventilation.

Hantavirus pulmonary syndrome (HPS): The strains of hantavirus are members of the *Bunyaviridae*. Infections usually occur after contact with an infected rodent or aerosolized rodent urine. Recent outbreaks in the southwestern U.S. led to lung failure as capillaries became leaky and fluid filled the air spaces.

Fungal Diseases

Histoplasmosis: The fungus *H. capsulatum* is the etiologic agent for this most common fungal respiratory disease. Aerosolized bird or bat feces contain fungal spores that are inhaled. The acute disease can be asymptomatic or symptomatic pneumonia. Without recovery, the chronic form leads to progressive pulmonary disease.

Blastomycosis: Inhalation of *B. dermatitidis* spores leads to an infection of the lungs and produces bronchopneumonia. The disease can spread to other parts of the body, including the skin where papules or pustules form.

Coccidioidomycosis (valley fever): Caused by *C. immitis,* the airborne conidia (arthrospores) are inhaled. In the bronchioles and alveoli, the spores form thick-walled spherules. Usually, the acute primary disease is without symptoms. This rare, progressive disease, usually in people with a weakened immune system, produces abscesses throughout the body that can be life threatening.

***Pneumocystis* pneumonia (PCP):** This opportunistic infection by *Pneumocystis jiroveci* (formerly *Pneumocystis carinii*) comes from the fungus already in the lungs as part of the human microbiota. Lung invasion leads to microbe and exudate-filled alveoli, resulting in a loss of gas exchange. PCP occurs in some 90% of HIV-infected patients.

Treatment

Serious cases of influenza illness can be treated with rimantadine. Otherwise, drugs to relieve the symptoms are the only recourse. For severe cases of RS disease, ribavirin is useful. No specific treatment is available for SARS or HPS.

Progressive disease cases of coccidioidomycosis may require hospitalization and perhaps lobectomy. The antifungal drug amphotericin B or fluconazole suppresses infection. Histoplasmosis and blastomycosis respond favorably to amphotericin B or oral itraconazole. Treatment of PCP involves trimethoprim-sulfamethoxazole, which has very toxic side effects.

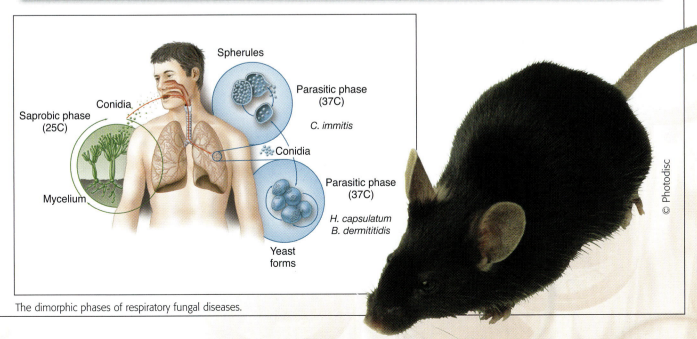

The dimorphic phases of respiratory fungal diseases.

Bacterial Infections

The digestive system consists of the mouth, salivary glands, pharynx, esophagus, stomach, small and large intestines, liver, gall bladder, and pancreas. Since part of this tract is a tube running through the body from mouth to anus, these areas are colonized by many normal microbiota species.

The survey of infections and diseases associated with the digestive tract will start with the mouth (oral cavity). Ingestion of contaminated food or water can lead to an inflammation of the stomach or intestinal mucosa. Pathogens in the food or water, or their **enterotoxins,** are to blame. Infections of the oral cavity and digestive system can be separated by their clinical **syndromes** (see opposite page).

Gingivitis: This extremely common inflammation of the gums often results from inadequate tooth brushing and flossing.

Periodontitis: This severe form of gingivitis extends into the supporting structures of the tooth and is a major cause of adult tooth loss.

Gastritis: This is an infection or inflammation of the stomach lining.

Gastroenteritis: An infection of the stomach or intestinal mucosa. Although the infections can be quite uncomfortable, gastroenteritis is rarely life threatening in adults.

Colitis: An infection of the colon (large intestine); also referred to as **dysentery,** it often results from damage to the intestinal wall.

Hepatitis: Damage to the liver can be the result of infectious microbes, especially viruses (see next unit).

Signs and Symptoms

Trench mouth: Gum pain, fever, occasional fatigue, foul breath, and bleeding gums.

Campylobacteriosis: Bloody or watery diarrhea, abdominal pain, cramps, and fever.

Bacterial gastroenteritis: Diarrhea, nausea, vomiting, abdominal cramps, and loss of appetite; fever, aching muscles, and extreme exhaustion may occur.

Cholera: Vomiting, painless, but severe watery diarrhea.

Traveler's diarrhea: Diarrhea, nausea, vomiting.

Typhoid fever: Prolonged fever, abdominal pain, and a rash (rose spots).

Dental caries: Food- or water-induced tooth pain if decay has reached dentin; lingering pain if decay enters the pulp.

Peptic ulcer: Burning, aching soreness with steady and often severe pain.

Salmonellosis: Nausea, cramping, abdominal pain, watery diarrhea, fever, and possibly vomiting after 6-48 hours.

Shigellosis: Abdominal pain, fever, watery stool with mucus and blood, dehydration.

Hemorrhagic colitis: Severe abdominal cramps, watery diarrhea; HUS complications: fatigue, weakness, sudden kidney failure.

Staphylococcal food poisoning: Severe nausea, vomiting after 1-6 hours; may include abdominal pain, headache, and fever.

Common Clinical Conditions

Bacterial diseases of the digestive system are outlined below.

Oral Diseases

Dental caries: *Streptococcus mutans* is a major **cariogenic** bacterium where decay into the pulp results in a painful cavity (caries).

Trench mouth: This gum inflammation is common among young adults not brushing their teeth. Also called **acute necrotizing ulcerative gingivitis** (**ANUG**), if left untreated, the infection can lead to bone and tooth loss.

Gastritis

Peptic ulcer: *Helicobacter pylori* survives the stomach's acid conditions and disrupts the mucosal lining. Inflammation can lead to peptic ulcer formation.

Gastroenteritis

Staphylococcal food poisoning: The illness (intoxication) comes from ingesting *S. aureus* toxins in contaminated food.

Typhoid (enteric) fever: *Salmonella typhi* is the most virulent serovar. Endotoxin release brings about the disease. After recovery, persons may remain carriers.

Salmonellosis: The intensity of *Salmonella* food poisoning depends on the number of bacteria ingested in contaminated products such as poultry.

Shigellosis: Food contaminated with *Shigella dysenteriae* gives rise to dysentery through the production of an enterotoxin.

Cholera: In the small intestine, an enterotoxin of *Vibrio cholerae* interferes with water and electrolyte reabsorption. Severe dehydration and death can occur without treatment.

Traveler's diarrhea: There are five pathogenic groups of *Escherichia coli.* The **enterotoxigenic *E. coli*** (**ETEC**) produce enterotoxins in the small intestine.

Hemorrhagic colitis: Toxins from *E. coli* O157:H7 damage the lining of the colon. Complications can lead to **hemorrhagic uremic syndrome (HUS)**.

Campylobacteriosis: *Campylobacter jejuni* invades and damages the mucosal surfaces of the small intestine and colon.

Other Forms of Bacterial Gastroenteritis

Several additional bacteria can cause bacterial gastroenteritis. These include:

- ***Clostridium perfringens:*** An obligate anaerobe transmitted through contaminated meat during slaughter. Enterotoxins produce abdominal cramps and diarrhea only.

- ***Vibrio parahaemolyticus*** and ***Vibrio vulnificus:*** These halophiles are associated with contaminated shellfish and produce little nausea or vomiting.

- ***Bacillus cereus:*** Can cause food poisoning by remaining viable in some improperly prepared oriental rice dishes where enterotoxins produce the characteristic symptoms.

- ***Yersinia enterocolitica:*** Often inhabits domestic animals. Invasion of cells of the small intestine causes gastroenteritis.

Treatment

For most cases of gastroenteritis caused by bacteria, antibiotics either are not given, or are given to shorten the duration of the illness and decrease the likelihood of transmission

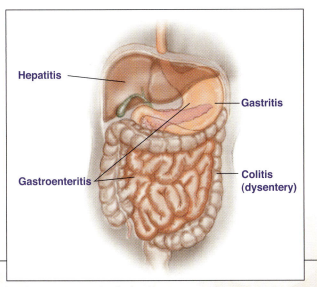

Syndromes associated with the digestive system.

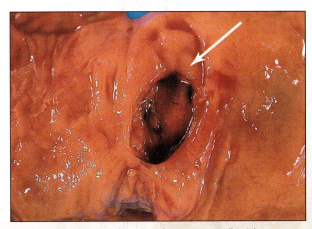

An ulcer in the lining of the stomach. (© Dr. E. Walker/Photo Researchers)

Fungal Intoxications and Viral Infections

Besides bacteria, other microbes or their products can affect the digestive system. Viruses are more common than bacteria in causing **gastroenteritis**. Any one of the five hepatitis viruses can cause **hepatitis**. An **acute** infection produces a sudden but short inflammatory period of a few weeks, while a **chronic** infection can persist for six months or longer.

Although not classified as disease, toxigenic molds produce **mycotoxins** and some mushrooms produce toxins that can have a variety of effects (**mycotoxicoses**) on humans, including digestive disorders. Intoxications due to mycotoxin poisoning arise from ingestion of the mycotoxin in food or ingesting poisonous mushrooms.

Aflatoxins: These mycotoxins are produced by a common mold, *Aspergillus flavus*. The toxins can cause liver damage and are potent carcinogens (cancer-causing agents) that can cause hepatomas (a type of liver cell tumor).

Ergot alkaloids: The fungus *Claviceps purpurea* can infect cereal grains (rye and wheat), during which time the mycotoxin is produced. Ergot poisoning (**ergotism**) arises from ingestion of these contaminated grain products. The toxin restricts blood flow at the limb extremities and causes hallucinations (psychotropic effects).

Mushroom poisoning: Some mushrooms, including several species of the genus *Amanita* (see opposite page) produce very dangerous toxins (phalloidin and amanitin) that can cause liver damage. Without treatment, about 50% of individuals who ingest phalloidin die in 5 to 8 days.

Signs and Symptoms

Hepatitis A: Often asymptomatic; anorexia, nausea, vomiting, fever, jaundice, and dark urine may occur.

Hepatitis B: Anorexia, fatigue, joint pain, itchy red hives (wheals) on skin; jaundice, and dark urine may occur.

Hepatitis C: Often asymptomatic; anorexia, nausea, vomiting, fever, jaundice, and dark urine may occur.

Enterovirus gastroenteritis: Watery diarrhea, vomiting, and fever.

Mumps: Chills, headache, malaise, and fever.

Hepatitis D: Occurs as a co-infection with hepatitis B.

Hepatitis E: Often asymptomatic; anorexia, nausea, vomiting, fever, jaundice, and dark urine may occur.

Rotavirus gastroenteritis: Fever, diarrhea, vomiting, abdominal cramps, nausea, and headache.

Norovirus gastroenteritis: Fever, watery diarrhea, vomiting, abdominal cramps, and nausea.

Other Forms of Hepatitis
Hepatitis also can result from other infections, including infectious mononucleosis, cytomegalovirus disease, and yellow fever. Claims of the existence of a hepatitis G virus have been withdrawn.

Viral diseases of the digestive system are outlined below.

Mumps: A disease of the oral cavity, the mumps virus (*Paramyxoviridae*) is transmitted by saliva and spreads to the salivary glands from the blood.

Viral Gastroenteritis

Rotavirus gastroenteritis: The human rotaviruses (*Reoviridae*) are spread hand to mouth among children from contaminated surfaces where children play. These viruses account for about 50% of all childhood hospitalizations for dehydration in the United States. Dehydration is more likely in infants, and they can become listless and lethargic.

Norovirus gastroenteritis: The highly contagious noroviruses (formerly called the Norwalk-like viruses; *Caliciviridae*) are spread via the fecal-oral route.

Enterovirus gastroenteritis: Two viruses among the nonpolio enteroviruses (*Picornaviridae*) are associated with gastroenteritis:

- The coxsackie A and coxsackie B viruses produce symptoms that usually resolve spontaneously.
- Many echovirus strains also produce symptoms without long-term complications.

Hepatitis

Hepatitis A (infectious hepatitis): The hepatitis A virus (HAV) has a short incubation and is spread via the fecal-oral route. Waterborne and foodborne epidemics are common in developing nations. Affected individuals usually recover completely from the acute infection.

Hepatitis B (serum hepatitis): The hepatitis B virus (HBV) has a long incubation and is most often transmitted by contaminated needles, through human secretions, or intimate sexual contact. In 5% to 10% of adult chronic cases (70% to 90% of infants), cirrhosis (destruction of liver tissue) or liver cancer (hepatocellular carcinoma) can develop.

Hepatitis C: The acute illness caused by the hepatitis C virus (HCV) usually is transmitted via the sharing of contaminated needles and in about 75% of cases, the disease becomes chronic and such affected individuals remain infectious. About 20% of affected individuals develop cirrhosis and liver cancer.

Hepatitis D: A hepatitis D virus infection occurs as a co-infection with acute or chronic hepatitis B infections (**delta hepatitis**) and often makes the HBV infection more severe. Like hepatitis B and C, transmission usually involves the sharing of contaminated needles.

Hepatitis E: The hepatitis E virus usually is spread via the fecal-oral route. This acute form of hepatitis occasionally causes waterborne epidemics, such as the recent epidemic in the Darfur region of Sudan. No chronic form of the disease is known to exist.

Treatment

For viral gastroenteritis, antibiotics and antiviral drugs are not prescribed. In addition, no specific drug therapy has been developed for hepatitis, except for hepatitis C. For individuals with chronic hepatitis C, a combination of interferon-alpha and ribavirin has been partially successful in stopping inflammation. In most cases, the disease recurs if treatment is discontinued.

A poisonous *Amanita* mushroom. (©Photodisc)

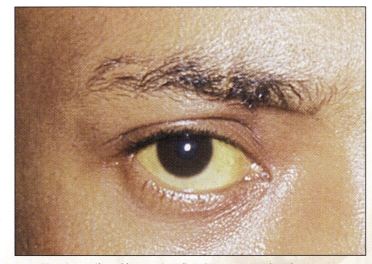

Hepatitis A is manifested here as jaundice. (Courtesy CDC/PHIL)

Parasitic Infections

A parasite is an organism that lives inside or on another organism, the **host**. They may be single-celled parasites—the **protozoa**—or multicellular parasites—the **helminths**. Many parasites enter the body through the mouth during the ingestion of contaminated food or water. Depending on the species, they may remain and reproduce in the intestines or spread to other organs of the body.

Besides bacteria and viruses, a few protozoa also cause **gastroenteritis**. These protozoa enter the body as resistant, infective **cysts** and, after completing their life cycle in the human digestive tract, more cysts are shed into the environment. During this process, a **trophozoite** (feeding) stage may be responsible for digestive system diseases.

The disease-causing helminths often are the least familiar infectious diseases in developed nations because helminthic parasites, being more common in areas where there is poor sanitation and hygiene, predominate in developing nations. Travelers should always follow the maxim: "Cook it, boil it, peel it, or forget it."

The life cycle of helminths requires both an **intermediate host** that harbors a larval form of the helminth and a **definitive host** that harbors the sexually mature adult. The disease-causing helminths include the flatworms (phylum Platyhelminthes) such as flukes and tapeworms, and the roundworms or nematodes (phylum Aschelminthes) that complete their entire life cycle in the human (definitive) host. The helminths may produce few symptoms because they have become so well adapted to living in their hosts.

Signs and Symptoms

Fluke infections: Diarrhea and intestinal blockage; liver damage.

Beef and pork tapeworm infections: Mild diarrhea and abdominal discomfort.

Ascariasis: Asymptomatic or abdominal pain with heavy worm loads.

Cryptosporidiosis: Profuse, chronic diarrhea, and abdominal cramps.

Whipworm infection: Abdominal pain and anemia.

Giardiasis: Nausea, foul-smelling watery diarrhea, abdominal cramping, and flatulence.

Cyclosporiasis: Watery diarrhea, nausea, abdominal pain, and vomiting.

Hookworm disease: Anemia and lethargic behavior.

Hydatid disease: Asymptomatic.

Amoebiasis: Abdominal pain and weight loss; bloody diarrhea in severe cases.

Pinworm infection: Diarrhea and anal itching.

Other Helminth Diseases of the Digestive System

Additional human parasites include *Diphyllobothrium latum* (fish tapeworm disease) [FOM p. 639], *Hymenolepis nana* (dwarf tapeworm disease) [FOM p. 640], and *Strongyloides stercoralis* (roundworm disease, strongloidiasis).

Protozoal and helminthic diseases of the digestive system are described below.

Protozoal Diseases

Amoebiasis: *Entamoeba histolytica* causes an inflammation of the colon. Ingested cysts hatch out trophozoites that produce ulcers on the surface of the intestinal lining.

Giardiasis: *Giardia lamblia* is transmitted via cyst-contaminated drinking water. The trophozoites infect and attach to the mucosa of the small intestine, triggering the disease.

Cryptosporidiosis: *Cryptosporidium parvum* is transmitted by fecally-contaminated water. Importantly, the cysts are not killed by normal chlorination of drinking water or swimming pool water.

Cyclosporiasis: The causative agent, *Cyclospora cayetanensis,* has a long incubation period. Outbreaks have involved foods contaminated with the protozoan.

Helminthic Diseases

Fluke infections: *Fasciolopsis buski* and *Fasciola hepatica* infect the intestine (duodenum) and liver, respectively.

Beef and pork tapeworm infections: *Taenia saginata* (beef tapeworm) and *T. solium* (pork tapeworm) are transmitted to humans (definitive host) in contaminated raw or undercooked meat. *T. saginata* interferes with normal physiological processes in the small intestine.

Hydatid disease: The tapeworm *Echinococcus granulosus* infects the intestines of dogs and other carnivorous animals (definitive host). Human (intermediate host) infection usually arises from ingestion of feces-contaminated materials containing helminth eggs. In the liver or lungs, the eggs develop into fluid-filled cysts, which if ruptured, can cause anaphylactic shock.

Pinworm infection: Ingestion of eggs from the roundworm *Enterobius vermicularis* leads to colon infestations, especially in children. Migration of the female worm to the anus results in the deposition of eggs in a gelatinous material.

Whipworm infection: Larvae of the roundworm *Trichuris trichiura* hatch in the small intestine and migrate to the colon where they embed in the intestinal lining.

Ascariasis: Ingestion of eggs from *Ascaris lumbricoides* allows their development in the small intestine where the worms live on partially digested food.

Hookworm disease: Larvae of *Necator americanus* and *Ancylostoma duodenale* enter through the skin of the foot. After traveling through the blood and lungs, they are passed to the small intestine and feed on blood and tissue.

Treatment

Amoebiasis and giardiasis can be treated with antiprotozoan drugs, such as metronidazole. No effective treatment exists for cryptosporidiosis. Trimethoprim-sulfamethoxazole is effective against cyclosporiasis.

Intestinal and liver flukes as well as beef and pork tapeworms can be treated with praziquantal. Hydatid disease is treated with albendazole. Roundworm infections can be treated with mebendazole.

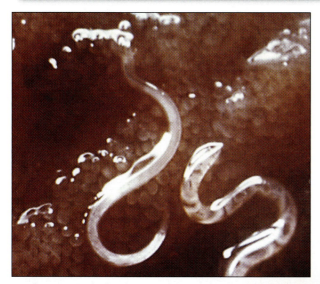

Ancylostoma attached to the intestinal mucosa. (Courtesy CDC/PHIL)

Giardia lamblia

Bacterial, Fungal, and Protozoal Infections

Background

The nervous system is made up of the central nervous system (CNS), which consists of the brain and spinal cord, and the peripheral nervous system, which consists of the nerves extending from the CNS.

Although rare, infections of the CNS by bacteria, fungi, protozoa, or viruses can be very serious, especially for infants and older adults. Infections may result in: (1) **meningitis,** an inflammation of the membranes surrounding the brain and spinal cord (the meninges). Such infections by bacteria or viruses cause swelling and increased pressure on the brain; (2) **encephalitis,** an inflammation of the actual brain tissue that is usually caused by certain groups of viruses (see next unit). Such infections produce behavioral changes; or (3) **myelitis,** an inflammation of the spinal cord by bacteria or viruses. Symptoms depend on the region and nerves affected.

Infections of the CNS usually are the result of a complication from another infection, including pneumonia, sinusitis, or otitis media. The agents of these diseases reach the brain via the blood, or from a sinus or middle ear infection. Infections also can be the result of a skull injury (fracture), head wound, or surgical procedure that exposes the CNS.

Although there is a large group of infectious bacteria that can cause meningitis, *Neisseria meningitidis* and *Streptococcus pneumoniae* are responsible for most cases of **acute bacterial meningitis.**

On rare occasions, fungi and protozoa infect the CNS. The fungus *Cryptococcus* can cause a type of **chronic meningitis,** while certain species of *Trypanosoma* are responsible for African sleeping sickness.

Signs and Symptoms

Cryptococcosis: Headache, dizziness, vertigo, and stiff neck.

Primary amoebic meningoencephalitis (PAM): Piercing headache, fever, delirium, and stiff neck.

Botulism: Flaccid paralysis, blurred vision, difficulty swallowing, weakness and fatigue.

Leprosy: Loss of skin pigment and sensation in tuberculoid form; deformity and destruction of tissues in the lepromatous form.

Acute bacterial meningitis: Severe headache, stiff neck, fever, and vomiting.

African trypanosomiasis: Fever, headache; later symptoms of lethargy, weakness, and anemia.

Tetanus: Abnormal muscle contraction, causing the back to arch and the mouth to remain closed; difficulty swallowing.

Infectious diseases of the nervous system are discussed below.

Acute Bacterial Meningitis

Meningococcal meningitis: The infection by *N. meningitidis* usually affects immunocompromised individuals, leading to bacteremia and meningitis. Symptoms are associated with the endotoxin produced.

Hemophilus meningitis: Infection by *H. influenzae* (type B) is preceded by a viral infection of the respiratory tract, and then infection of the meninges occurs.

Pneumococcal meningitis: The infection by *S. pneumoniae* results in the highest rate of meningitis.

Listeric meningitis (listeriosis): *Listeria monocytogenes* preferentially infects the CNS of immunocompromised individuals. Infection in pregnant women can cause high rates of spontaneous abortions or stillbirths.

Other Forms of Meningitis

Cryptococcal meningitis (cryptococcosis): Infection by the yeast-like fungus *Cryptococcus neoformans* affects persons with an impaired immune system. The brain and meninges are most affected.

Primary amoebic meningoencephalitis (PAM): An acute disease caused by the protozoan *Naegleria fowleri* or a more chronic illness due to *Acanthamoeba*. Most patients die from the infection.

Other Bacterial Diseases

Botulism: This severe form of poisoning is caused by the neurotoxins produced by *Clostridium botulinum.* Transmission usually is the result of ingesting contaminated foods. Nerve transmission is blocked due to toxin inhibition of acetylcholine release at nerve endings.

Tetanus: Also called lockjaw, this systemic disease is due to a neurotoxin produced by *Clostridium tetani,* which blocks nerve transmission inhibitors.

Leprosy: *Mycobacterium leprae* produces a slowly progressive systemic infection (also called **Hansen's disease**). Bacterial multiplication in the peripheral nerves and skin macrophages leads to a tuberculoid or lepromatous form, the latter producing disfiguring skin lesions (**lepromas**) over the body.

Other Protozoal Diseases

African trypanosomiasis: Transmitted by the bite of the tsetse fly, either of the two varieties of *Trypanosoma brucei* multiplies in the blood and infects the brain. One characteristic is lethargy, and thus the common term is **African sleeping sickness** for this disease.

Treatment

Bacterial meningitis requires quick action and therapy to prevent disabilities and death. Antibiotic treatment (ampicillin, cephalosporins) can be used on susceptible bacteria. Persons infected with *C. neoformans* can be treated effectively with amphotericin B. There is no treatment for PAM.

Treatment for botulism requires stomach pumping and antitoxin therapy. Tetanus requires immune globulin. Leprosy can be controlled with rifampin and dapsone. For the blood/lymph stage of African trypanosomiasis, suramin can be used. For the neurological stage, melarsoprol is the drug of choice.

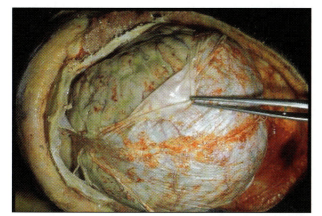

Pneumococcal meningitis at autopsy. (Courtesy CDC/Dr. Edwin P. Ewing, Jr.)

Trypanosoma brucei

Viral and Viral-like Infections

Background

There are several viral diseases of the CNS. These viruses enter the CNS from the blood or peripheral axons. Most often, the viruses cause **meningitis, encephalitis,** or **myelitis.** Such viral inflammations, especially encephalitis, can produce devastating and sometimes fatal results.

Some infections occur in epidemics, such as the encephalopathies caused by the echovirus or Coxsackie virus. In addition, the herpesviruses, mumps, and chickenpox viruses can cause isolated cases of encephalitis or myelitis. Infection by the human immunodeficiency virus (HIV) can produce a chronic infection of the brain without characteristic acute encephalitis. Such situations are referred to as **AIDS dementia.**

Many of the encephalopathies caused by the **arboviruses** (*ar*thropod-*bo*rne viruses) are spread by mosquitoes, ticks, or other arthro-pods. The rabies virus is transmitted by the bite of a rabid animal, while lymphocytic choriomeningitis (LCM) is spread by the exposure to viruses in rodent feces or urine.

A term often used to describe meningitis when no bacterial cause can be found is **aseptic meningitis.** Most often, a virus is the cause of an acute or chronic form and can be referred to as **viral meningitis.** Such cases usually are mild and rarely life-threatening.

There also are diseases caused by virus-like agents, specifically **prion disease,** such as Creutzfeldt-Jacob disease (CJD) and Gertsmann-Sträussler-Scheinker disease. The agents for these diseases, abnormal **prion proteins,** may be acquired from an external source, such as prion-contaminated beef (variant CJD), or the disease may occur from an inherent genetic error.

Signs and Symptoms

Prion disease: Progressive deterioration of muscle function, muscle twitching, and staggered walking.

Lymphocytic choriomeningitis (LCM): Headache, fever, malaise, drowsiness, and stupor.

Rabies: Initially a headache, fever, and abnormal sensations at bite site; later CNS involvement with periods of alert and aggressive behavior followed by muscle paralysis and hydrophobia.

Arboviral encephalitis: Sudden high fever, severe headache, drowsiness, stiff neck; personality changes, seizures, and coma develop rapidly.

Progressive multifocal leukoencephalopathy (PML): Paralysis on one side of the body; followed by a decline in mental function, and dementia.

Poliomyelitis: Nausea, vomiting, and cramps; active cases in CNS show loss of motor function and severe paralysis.

Common Clinical Conditions

Viral and viral-like diseases of the CNS are outlined below.

Viral Diseases

Rabies: The rabies virus (*Rhabdoviridae*) is transmitted to humans in the saliva of a rabid animal bite. The virus slowly spreads through the blood to the CNS. Once in the CNS, neuronal damage causes fatal encephalitis.

Poliomyelitis: Spread through poor hygiene or inhalation from infected individuals through pharyngeal secretions, the poliovirus (*Picornaviridae*) multiplies in the gastrointestinal tract. **Paralytic polio** (myelitis) involves virus spread to the motor neurons of the CNS. Death occurs when there is a loss of respiratory control.

Lymphocytic choriomeningitis (LCM): The LCM virus (*Arenaviridae*) produces a mild, flu-like illness often characterized as **aseptic meningitis.** The virus infects the lymphocytes of the meninges. The mortality rate is low.

Progressive multifocal leukoencephalopathy (PML): This infection by the JC virus (*Polyomaviridae*) primarily affects immunocompromised individuals. Death most often occurs within 1 to 6 months after symptoms appear.

Arboviral Encephalitis

An infected mosquito, carrying the virus from a reservoir (horses and birds) to human, introduces the virus into the blood during a blood meal. Primary encephalitis occurs when the virus infects neurons in the brain, causing brain swelling and hemorrhage. Permanent nervous system damage may result. Several types of encephalitis caused by arboviruses occur in the United States.

St. Louis encephalitis (SLE): A member of the *Flaviviridae,* the virus most often causes disease in Texas and Midwestern states.

California and La Crosse encephalitis: Caused by members of the *Bunyaviridae,* these viruses mainly affect children in the Western states (California virus) and Midwestern states (La Crosse virus).

Eastern equine encephalitis (EEE): The virus, a member of the *Togaviridae,* mainly affects young children and people over 55 years old in the Eastern U.S.

Western equine encephalitis (WEE): Also caused by a virus in the *Togaviridae,* the virus can affect any individual, but predominantly infants. The disease occurs throughout the U.S. WEE is less likely to be fatal than EEE.

Venezuelan equine encephalitis (VEE): Another membrane of the *Togaviridae,* the virus primarily infects young children and the elderly in Florida and Texas.

West Nile encephalitis (WNE): This virus is a member of the *Flaviviridae* and is closely related to the SLE virus. Several species of birds are the viral host. Encephalitis mainly affects the elderly. About 10% of such individuals succumb to the disease.

Disease Caused by Virus-Like Agents

Prion disease: A family of related diseases caused by a specific protein called a **prion** that attacks the CNS. Most patients survive for 3 to 12 months after disease symptoms are diagnosed. Pneumonia is most often the cause of death.

✚ Treatment

For rabies, the bite wound should be thoroughly cleaned with soap and water. The individual then should be given post-exposure immunization (rabies immunoglobulin). There is no treatment for paralytic polio, LCM, PML, or arboviral encephalitis.

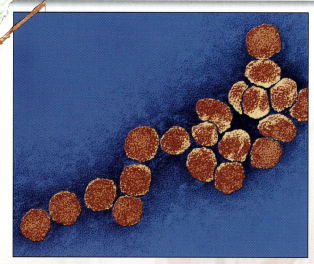

Transmission electron micrograph of West Nile virus. (©Science VU/Visuals Unlimited)

Bacterial Infections

The cardiovascular system consists of the heart, blood, and blood vessels, while the lymphatic system is made up of the lymph, lymph vessels and nodes, and the lymphoid organs.

Several species of bacteria cause disease in the cardiovascular or lymphatic systems, or are spread via these systems. Among them are bacteria that cause disease via contact transmission. Once in the body, they may spread via the blood or lymph to other body parts. Other bacterial diseases arise from animal bites. These may be infections resulting from bites by domestic animals (dogs and cats) or from bites by infected fleas, lice, or ticks. In fact, there is a whole group of tickborne bacterial diseases.

There are different outcomes if bacteria enter the bloodstream:

Bacteremia: This is the presence of live bacteria in the blood. Temporary bacteremia can occur from dental or surgical procedures. Simple tooth brushing that causes the gums to bleed can introduce bacteria into the bloodstream. These conditions usually are not life-threatening.

Septicemia: Less common is the presence (infection) of persistent bacteria that divide in the blood. The condition though is more serious than bacteremia—and often fatal. **Sepsis** can result from another infection in the body that spreads through the bloodstream and can give rise to a systemic disease.

Septic shock: This life-threatening condition occurs if sepsis causes a sudden lowering of blood pressure and organ malfunction. A combination of immune system proteins (cytokines) and bacterial toxins often triggers vascular collapse. Several species of bacteria, including those of *Staphylococcus, Streptococcus, Pseudomonas, Klebsiella,* and *Enterobacter,* can cause septic shock.

Signs and Symptoms

Relapsing fever: Periods of fever, chills, and headache separated by apparent recovery.

Rickettsial infections: Fever, malaise, severe headache.

Plague: Lymph node hemorrhage and buboes.

Anthrax: Inhalation: malaise, mild fever, dry cough, and blue coloration of neck and chest.

Cat-scratch disease: Rash, headache, fever, swollen lymph nodes.

Lyme disease: Stage 1: bull's-eye rash at bite site; Stage 2: facial palsy; Stage 3: chronic arthritis.

Brucellosis: High fever, backache, and joint pain.

Tularemia: Ulcer at infection site, swollen lymph nodes, and, if uncontrolled, septicemia.

Infective endocarditis: Fever, fatigue, anemia, and heart murmurs.

Rheumatic fever: Fever, joint pain, and rash.

Gas gangrene: Fever, pain, swelling, and blackening of the skin at wound site; foul odor.

Rat-bite fever: Recurrent fever, arthritic pain, rash.

Common Clinical Conditions

Bacterial diseases of the cardiovascular and lymphatic systems are outlined below.

Diseases of the Heart

Infective (bacterial) endocarditis: A rare infection of the endocardium or heart valves, *Staphylococcus* and *Streptococcus* clots can form emboli in distant organs. Untreated, endocarditis always is fatal.

Rheumatic fever: Complications from pharyngitis by *Streptococcus pyogenes* can seriously damage the heart valves, which can be fatal.

Contact Diseases

Brucellosis: Several species of *Brucella* cause the disease (also called **undulant fever**). The bacteria spread through the lymphatic system and into the blood.

Anthrax: **Exotoxins** produced by *Bacillus anthracis* cause an acute disease that can be spread through the blood causing multisystem malfunction.

Gas gangrene: The disease occurs if devitalized wounds or surgical sites become infected with *Clostridium perfringens.* Systemic effects and heart damage can result from toxin production.

Tularemia: *Francisella tularensis* causes ulceration at the infection site and enlarged lymph nodes that can become abscessed. Septicemia and death may occur.

Diseases from Animal Bites

Plague: Transmission of *Yersinia pestis* by a fleabite causes bloodborne bubonic and septicemic plague; pneumonic plague produces almost 100% mortality.

Lyme disease: Adult *Ixodes* ticks transmit the spirochete *Borrelia burgdorferi* to humans. Symptoms occur in three stages.

Relapsing fever: Other *Borrelia* spirochetes, carried by lice, are responsible for periods of fever followed by apparent recovery.

Cat-scratch disease: Cat bites can produce a benign infection in the lymph nodes of children. Either *Afipia felis* or *Bartonella henselae* are the cause.

Rat-bite fever: The genera *Streptobacillus* and *Spirillum* cause infection from bites by rats or other infected small rodents.

Rickettsial Diseases

Rocky Mountain spotted fever: *Rickettsia rickettsii* (tick-borne) causes a rash, headache, and high fever.

Epidemic typhus: *Rickettsia prowazekii* (louse-borne) produces sudden fever, rash, headache, and fatigue.

Endemic typhus: *Rickettsia typhi* (flea-borne) produces symptoms similar to, but milder than epidemic typhus.

Ehrlichiosis: *Ehrlichia chaffeensis* (tick-borne) causes abrupt fever, chills, headache, and malaise.

Treatment

Endocarditis can be treated with IV antibiotics, while pharyngitis requires prompt treatment with penicillin to prevent rheumatic fever. Early treatment of anthrax, and treatment of brucellosis and tularemia, with antibiotics can be successful. Penicillin for gas gangrene will stop the spreading, but surgical removal of infected tissue may be required.

Prompt treatment with streptomycin can decrease greatly the chance of death from plague. Likewise, early treatment with doxycycline is effective against Lyme disease and relapsing fever. Rat bite fever and cat-scratch disease are treatable with tetracycline. Doxycycline also is effective against any of the rickettsial infections.

Courtesy CDC/Michael L. Levin, Ph.D.

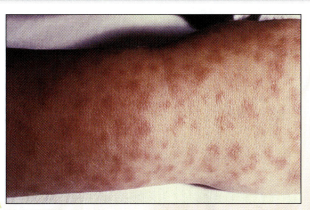

Characteristic rash of Rocky Mountain spotted fever.
(Courtesy CDC/PHIL)

Parasitic and Viral Infections

Background

Protozoal diseases of the cardiovascular and lymphatic systems can have profound affects on humans. Malaria represents one of the most devastating of all human diseases. Other protozoal diseases include leishmaniasis, Chagas' disease (American trypanosomiasis), and toxoplasmosis.

Several viruses can infect the cardiovascular and lymphatic systems. Some of the viral diseases are characterized by bleeding and are referred to as **hemorrhagic fevers**. Also included in this unit is the **human immunodeficiency virus (HIV)**, which is responsible for **HIV disease** and **acquired immunodeficiency syndrome (AIDS)**. Although the virus is transmitted sexually, HIV does not attack the reproductive system. Rather, it depletes T lymphocytes that are a major part of the lymphatic system. Without these cells, the immune system is almost helpless in fighting opportunistic infections arising from the HIV infection.

A person usually does not die from the HIV infection, but rather from secondary, opportunistic infections resulting from immune system destruction (AIDS). The typical sequence of patient conditions is:

HIV disease: This is an acute HIV infection when HIV antibody appears; flu-like symptoms may be present but soon disappear.

Early HIV symptoms: The patient has a slight decrease in T-cell count and mild leukopenia with immune system dysfunction.

Advanced HIV symptoms: The patient has a more severe drop in T cells and moderate leukopenia with immune system dysfunction (yeast infections, diarrhea, weight loss, and bacterial infections are common).

AIDS: Individuals exhibit severe immune dysfunction (opportunistic infections, wasting disease, and lymphomas can occur). Kaposi's sarcoma and neurological problems often are evident and can appear with early to advanced HIV symptoms.

Signs and Symptoms

Viral hemorrhagic fevers: Bleeding, fever, vomiting, diarrhea; chest pain, and body aches (Lassa fever).

Infectious mononucleosis: Asymptomatic or fatigue, mild fever, sore throat, and enlarged lymph nodes.

Leishmaniasis: Visceral: fever, spleen and liver enlargement, weakness, and emaciation.

Schistosomiasis: As the worms multiply over several years, fever, fatigue, and an enlarged liver result.

Toxoplasmosis: Asymptomatic or chills, fever, fatigue, and headache.

Babesiosis: Anemia, piercing headache, and fever.

Dengue fever: High fever, muscle and joint pain, headache, and a rash; petechiae in hemorrhagic forms.

Malaria: Recurring cycles of chills and fever, anemia, headache, and delirium.

Chagas' disease: Fever, swelling at bite site, and liver and spleen enlargement.

Yellow fever: Fever, headache, nausea; followed by vomiting (blood), jaundice, nosebleeds, and bleeding from gums and the gastrointestinal tract.

Filariasis: Fever, chills, and swelling of the lymphatic system in groin or limbs.

Other Viral Fevers
There are many other viral fevers, including: Colorado tick fever, sandfly fever, Rift Valley fever, and other viral hemorrhagic fevers that are members of the *Bunyaviridae* or *Arenaviridae*.

Common Clinical Conditions

Parasitic and viral diseases of the cardiovascular and lymphatic systems are outlined below.

Parasitic Diseases

Chagas' disease: Once in the blood, the protozoan *Trypanosoma cruzi* can cause a chronic infection that affects the heart muscle.

Leishmaniasis: The more serious visceral form caused by the protozoan *Leishmania* infects white blood cells and lymph nodes.

Toxoplasmosis: Ingested cysts of *Toxoplasma gondii* can pass across the placenta, congenitally affecting the fetus.

Malaria: Once in the blood, the *Plasmodium* protozoa infect the liver and erythrocytes. Organ damage can lead to a fatal drop in blood pressure.

Babesiosis: The protozoan *Babesia microti* produces a mild disease that, in healthy individuals, disappears on its own.

Schistosomiasis (bilharzia): After migration to the blood, the *Schistosoma* fluke can cause cirrhosis of the lungs and liver; egg release can produce an inflammation that blocks veins in the bladder, liver, or intestines.

Filariasis: Larvae of *Wuchereria bancrofti* mature into adult worms and infect the lymph glands and vessels. Over years, the roundworms block lymph channels, causing a swelling (**elephantiasis**) of the limbs or scotum.

Viral Diseases

Yellow fever: The yellow fever virus (*Flaviviridae*) infects many tissues throughout the body, especially the liver.

Dengue fever: In the lymphatic system, the dengue fever virus (*Flaviviridae*) produces sudden, but mild symptoms. However, the disease can progress to a serious hemorrhagic fever with subcutaneous bleeding (**petechiae**) or even to toxic shock.

Epstein Barr virus (EBV) infections: EBV (*Herpesviridae*) causes lymphocyte proliferation that results in **infectious mononucleosis.** On rare occasions, EBV contributes to the development of cancers, such as **Burkitt's lymphoma.**

Viral Hemorrhagic Fevers

Lassa fever: These *Arenaviridae* viruses are found mainly in West Africa and are transmitted from infected rodents. Death can occur from blood leakage and shock.

Marburg and Ebola hemorrhagic fevers: These members of the *Filoviridae* are transmitted by blood or body tissue exposure. Both infections often are fatal.

☤ Treatment

Nifurtimox is partially successful in treating Chagas' disease, stibogluconate for leishmaniasis, and sulfonamide drugs for toxoplasmosis. Malaria can be treated with chloroquine and mefloquine. Schistosomiasis can be treated with praziquantal and albendazole, and diethylcarbamazine is prescribed for filariasis.

Aspirin or ibuprofen relieves the fever and pain of infectious mononucleosis. There is no effective treatment for HIV disease/AIDS, dengue fever, or the viral hemorrhagic fevers, although ribavirin can reduce the death rate from Lassa fever.

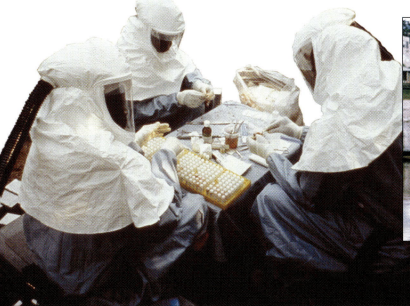

Testing for Ebola during a 1995 Zaire outbreak. (Courtesy CDC/PHIL)

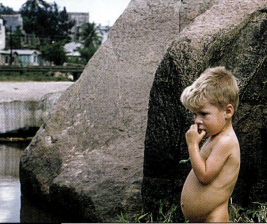

This Puerto Rican boy's swollen abdomen is due to schistosomiasis. (Courtesy CDC/PHIL)

Infections of the Reproductive System

Background

The reproductive system in males and females is involved with the production of gametes—sperm and eggs, respectively. These tracts normally are sterile, except for the female vagina that is colonized by several types of resident microbiota.

Several bacterial pathogens can infect the female reproductive tract, causing **vaginitis** (inflammation of the vagina), **endometriosis** (inflammation of the uterine lining), **salpingitis** (inflammation of the fallopian tubes), or **oophoritis** (infection of the ovaries). Infection of the male reproductive tract often causes **urethritis**.

Besides the fungi and protozoa that infect the human reproductive tract, a few viruses also are associated with the system. These viral diseases can be quite serious, often leading to cancer.

Most reproductive system pathogens cause **sexually transmitted diseases** (STDs). Some STDs are systemic, while others remain associated with the female or male reproductive system. Besides the diseases described under the common clinical conditions, a few other nonsexually-transmitted diseases can be transmitted through sexual contact. Except for hepatisis B and C, symptoms typically involve diarrhea, fever, abdominal pain, nausea, vomiting, and jaundice from oral contact during sex. These diseases include:

Bacterial: Campylobacteriosis; salmonellosis; shigellosis.

Viral: Cytomegalovirus (CMV) infection; hepatitis A, B, and C.

Protozoal: Amoebiasis; Giardiasis.

Signs and Symptoms

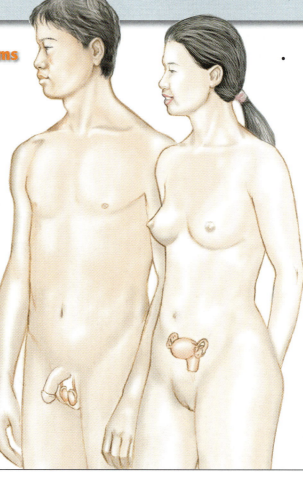

Genital warts: Warts on external genitalia.

Genital herpes: Fluid-filled vesicles appear that rupture into painful ulcers.

Candidiasis: White vaginal discharge and vaginal itching.

Chanchroid: Soft chancre on genitals and swollen lymph nodes.

Nongonococcal urethritis (NGU): Asymptomatic or painful urination and mucoid discharge.

Lymphogranuloma venereum (LGV): Sore on genitals, fever, headache followed 1 to 8 weeks later by swollen lymph nodes that become buboes. Healing produces scars that can block lymphatic vessels.

Syphilis: Primary: chancre lasting 2 to 3 weeks; Secondary: skin rash, fatigue, fever, enlarged lymph nodes occurring 8 to 11 weeks after infection; Tertiary: neurological symptoms and gummas occur 5 to 40 years after infection.

Gonorrhea: Male: urethra inflammation, painful, burning urination, and pus discharge; Female: asymptomatic or burning urination, abdominal/pelvic pain, and pus discharge.

Pelvic inflammatory disease (PID): Fever, abnormal vaginal bleeding, and severe lower abdominal pain.

Trichomoniasis: Female: asymptomatic or itching, vaginal stinging, burning urination, and yellowish discharge; Male: asymptomatic or urethritis and burning urination.

Granuloma inguinale: Genital lesions that ulcerate; loss of skin pigmentation.

Toxic shock syndrome: High fever, vomiting, diarrhea, hypotension, and red skin rash.

Diseases of the reproductive system are outlined below.

Bacterial Diseases

Toxic shock syndrome (TSS): The toxin produced by some strains of *Staphylococcus aureus* is absorbed through the vaginal mucosa.

Sexually Transmitted Diseases (STDs)

Syphilis: Infection with the spirochete *Treponema pallidum* produces a **chancre** at the contact site. Without treatment, the infection may spread, producing a red rash. Many years later, the infection can involve several body systems. Blindness, insanity, and death can occur.

Gonorrhea: *Neisseria gonorrhoeae* causes an acute, infectious disease that can penetrate the mucosa and subepithelial tissues. Untreated males and females may become sterile. Pregnant women with untreated gonorrhea can transmit the disease to the eyes of the newborn during delivery.

Nongonococcal urethritis (NGU): *Chlamydia trachomatis* and the mycoplasmas *Ureaplasma urealyticum* and *Mycoplasma hominis* infect the vagina in females and the urethra in females and males. "Chlamydia" (**chlamydial urethritis**) represents the most common NGU.

Pelvic inflammatory disease (PID): Bacterial infections of the female reproductive organs and abdominal cavity by *Chlamydia trachomatis, Neisseria gonorrhoeae,* or *Mycoplasma hominis* cause PID.

Chancroid: *Haemophilus ducreyi* forms a soft, painful chancre on the genitals. Without treatment, chancre healing can damage or scar the genitals.

Lymphogranuloma venereum (LGV): Caused by additional serotypes of *C. trachomatis,* the disease is rare, but left untreated it can cause permanent lymphatic or rectal blockage.

Granuloma inguinale: *Calymmatobacterium granulomatis* produces small vesicles on the genitals. The lesions spread to form granulomas that ulcerate.

Genital herpes: Herpes simplex virus type 2 (HSV-2) affects the external genitalia. Infection of a fetus (**neonatal herpes**) by an infected mother can be fatal to the fetus or cause encephalitis.

Genital warts: Sexually transmitted papillomaviruses can cause growths on the penis or vagina.

Candidiasis (vulvovaginitis): Infection by the fungus *Candida albicans* is one of the most common causes of vaginitis. An infection usually is opportunistic due to loss of resident microbiota through antibiotic use or immune suppression.

Trichomoniasis: The flagellated protozoan *Trichomonas vaginalis* can cause vaginitis or urethritis in females, urethritis or prostatitis (inflammation of the prostate) in males.

⚕ Treatment

Depending on the microbe causing vaginitis, antimicrobial agents can be used. TSS treatment involves using cephalosporins. PID also can be treated with cephalosporins and tetracycline. Gonorrhea and syphilis are treated with penicillin, doxycycline is used for NGU, and tetracycline is effective on chancroid and granuloma inguinale. For vaginal infections by *Candida*, nystatin and miconazole are effective. Oral acyclovir can be useful in treating primary infections of genital herpes. Metronidazole can be used for treatment of trichomoniasis.

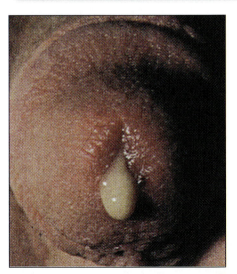

Purulent penile discharge due to gonorrhea. (© Visuals Unlimited)

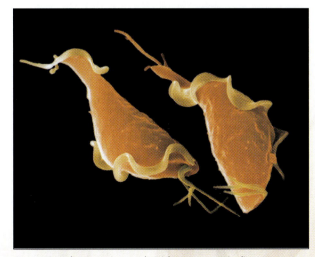

A scanning electron micrograph *Trichomonas vaginalis.*
(© Dr. David M. Phillips/Visuals Unlimited)

Infections of the Urinary System

Background

The urinary tract consists of the kidneys, two ureters, the urinary bladder, and the urethra. Because the outflow of urine from the bladder through the urethra is sterile, the urethra normally contains few resident microbes.

Several pathogens can cause infections and diseases when they invade the urinary tract. Such **urinary tract infections (UTIs)** are quite common; greater than 85% of UTIs are caused by bacteria from the intestine or vagina, resulting in an ascending infection. Infections also can be bloodborne and spread to the kidneys.

Although about 50% of UTIs are asymptomatic, UTIs can exhibit several clinical **syndromes**, depending on the anatomic system affected (see opposite page).

Urethritis: This is an infection of the urethra. Often painful urination (dysuria) and a mucoid discharge occur. The syndromes are usually the result of sexually-transmitted microbes.

Cystitis: This infection of the bladder usually is the result of a lower urinary tract infection involving enteric microbes. Besides dysuria and mucoid discharge, the urine is often cloudy and malodorous.

Pyelonephritis: A bacterial infection of the kidneys. Up to 50% of pregnant women suffering from acute pyelonephritis give birth to premature babies.

Glomerulonephritis: An infection of the glomeruli of the kidneys is the result of an immune complex hypersensitivity. Fever, high blood pressure, and protein and red blood cells in the urine are common.

Signs and Symptoms

Cystitis: Frequent urination and pain in the lower back.

Urethritis: Painful and frequent urination in females; urethral discharge (pus or mucus) in males.

Leptospirosis: Asymptomatic or abrupt onset of fever, headache, and muscle aches; nausea and vomiting; meningitis can occur in untreated cases.

Poststreptococcal glomerulonephritis: Edema, low urine volume, dark urine; red blood cells in urine; hypertension.

Common Clinical Conditions

Diseases of the urinary system are outlined below.

Bacterial Diseases

Leptospirosis: The spirochete *Leptospira interrogans* is transmitted to humans accidentally. Urine or urine-contaminated soil from an infected animal enters the mucous membranes of the eyes or nose, or through a skin abrasion. Infected individuals may be asymptomatic. Up to 10% of patients experience **Weil disease,** which occurs if the bacteria spread to the liver and kidneys.

Poststreptococcal glomerulonephritis: A postinfection complication of a childhood strep throat or skin infection by *Streptococcus pyogenes* can lead to autoimmune kidney damage. Recovery usually is complete.

Organisms Causing Urethritis

- *Neisseria gonorrhoeae,* the bacterium causing gonorrhea, can spread to the urethra during sexual intercourse.
- *Chlamydia trachomatis* is responsible for a form of urethritis called **nongonococcal urethritis (NGU).**
- Herpes simplex virus type 2 (HSV-2; *Herpesviridae*) may infect the urethra, causing painful urination and making urination difficult.
- *Trichomonas vaginalis* usually infects the vagina; however, the protozoan also can infect the bladder or urethra in women.

Organisms Causing Cystitis

- *Escherichia coli* is the most common cause of UTIs, such as cystitis. Several strains are uropathogenic and contain virulence factors, such as fimbriae, which attach to the epithelial cells lining the urinary tract. The bacterium also causes 90% of kidney infections (pyelonephritis).
- *Proteus mirabilis:* If a person has a kidney stone, this bacterium may be able to grow in the urine and infect the bladder.
- *Candida albicans:* This fungus causes **candidiasis** and most frequently infects individuals with a weakened immune system or who have a bladder catheter in place.
- *Schistosoma:* This flatworm (fluke) can infect the bladder, ureters, and kidneys causing kidney failure. Persistent infections of the bladder can lead to bladder cancer.

☤ Treatment

Leptospirosis can be treated with penicillin. Antibiotics usually are ineffective against poststreptococcal glomerulonephritis.

UTI treatment is designed to remove the pathogen from the urine and urinary tissues. Cases of urethritis due to (1) bacterial infections can be treated with antibiotics, (2) fungal infections are given antifungal drugs, or (3) HSV-2 viral infections can be treated with acyclovir.

Cystitis can be treated effectively with sulfa drugs, while pyelonephritis requires intravenous injections of cephalosporins. Prevention of cystitis and pyelonephritis require proper urinary tract hygiene.

Nosocomial cases of UTIs often come from bacteria transmitted from catheterized patient to catheterized patient on the hands of medical personnel. Proper hand washing is important to prevent such transmission.

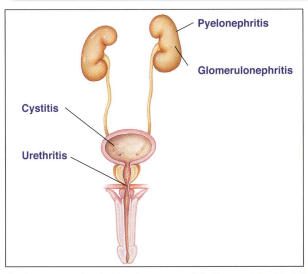

Syndromes associated with urinary system infections.

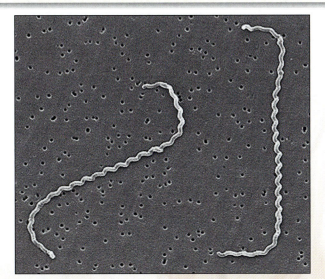

A scanning electron micrograph of *Leptospira interrogans.* (Courtesy CDC/NCID/Rob Weyant)